stude

WO OOK

AS Physics

Mechanics & Bulk Properties of Solids

Sophie Jacobi

Philip Allan Updates, an imprint of Hodder Education, part of Hachette Livre UK, Market Place, Deddington, Oxfordshire OX15 0SE

Orders
Bookpoint Ltd, 130 Milton Park, Abingdon, Oxfordshire OX14 4SB
tel: 01235 827720 fax: 01235 400454
e-mail: uk.orders@bookpoint.co.uk

Lines are open 9.00 a.m.–5.00 p.m., Monday to Saturday, with a 24-hour message answering service. You can also order through the Philip Allan Updates website: www.philipallan.co.uk

ISBN 978-1-84489-484-0

First printed 2008

Impression number 5 4 3 2 1

Year 2013 2012 2011 2010 2009 2008

Printed in Spain

With thanks to Ibstock Place School

Environmental information
Hachette Livre UK's policy is to use papers that are natural, renewable and recyclable products and made from wood grown in sustainable forests. The logging and manufacturing processes are expected to conform to the environmental regulations of the country of origin.

Introduction

This workbook is designed to be used in conjunction with your chosen textbook and class notes to consolidate your understanding. Your teacher may set you questions from this workbook as homework exercises, or you may use it for revision purposes. Take time to read the notes at the start of each topic; ensure that you understand everything and consult a more detailed source or ask your teacher if you find anything you are unsure about. Remember that every specification is slightly different, so there may be some material here that you do not need for your course, or some detail that is missing.

There are four main types of question in this book:

- routine questions and calculations to **practise** using formulae
- questions to make you **think** about what you have learned — these types of question include writing definitions, filling in summary tables, writing possible questions given the answers and spotting mistakes in sentences
- questions to **push** you — some of the questions in this workbook are harder than those you would get in an exam; force yourself to approach them logically, re-reading the source material if necessary
- **synoptic** questions to help you to practise linking different areas of the course

It is strongly recommended that, whenever you are completing practice questions, you have with you a copy of the data sheet and the calculator that you will use in the final exams. In this way, you can get used to finding your way around the data and formulae you will be given in the exams, as well as becoming familiar with using your calculator. This will make you feel much more comfortable during the exam.

It is important to develop, as early as you can, a good technique for presenting numerical solutions; be especially careful to show all your working and add words where they help to clarify what you are doing. Take note of prefixes such as 'nano' and 'micro', and do check your answer when you have finished — is it sensible? Remember that you can lose up to 1 mark per question for missing or incorrect units and a further mark per paper for errors in rounding (three significant figures is almost always appropriate) and in using standard form.

Topic 1 Vectors

Scalars and vectors

A scalar is a quantity that has magnitude (size) only.
Examples of scalars include distance, speed, time and mass. These values do not need to be qualified by a direction; for example, an object does not have a mass of 5 kg north, nor do we specify a time interval as 37 s to the left. Scalars can therefore be represented by a number (and unit) alone and can be added by simple arithmetic.

A vector quantity has both magnitude and direction.
Examples of vectors include displacement, velocity, acceleration and forces such as weight. Vectors can be represented by arrows, where the length of the arrow denotes the vector's magnitude.

A **negative sign** in front of a vector can be used to indicate a reversal of direction. For example, if (+)5 N means a force of 5 N acting to the right, then –7 N denotes a force of 7 N acting to the left. A special use of the negative sign is made when talking about acceleration: in this case, a negative acceleration denotes deceleration. For example, an acceleration of $5\,m\,s^{-2}$ means that, every second, the velocity of an object is increasing by $5\,m\,s^{-1}$, i.e. it is speeding up by $5\,m\,s^{-1}$ per second; an acceleration of $-5\,m\,s^{-2}$ indicates that the object is slowing down at the same rate.

Adding vectors

The **resultant** of a system of vectors is a single vector that represents the whole system; it is found by adding the individual vectors. The direction of each vector has to be taken into account when adding vectors.

Parallel vectors

The resultant of a system of vectors acting parallel or anti-parallel to each other can be found by simply adding or subtracting them, respectively.

Key point: You must always state the direction that a vector is acting in.

Vector triangles

Two coplanar vectors can be added by drawing a scale triangle of vectors as follows:

1. Represent each of the forces using an arrow, drawn to scale (e.g. 1 cm to 2 N) and with the arrow pointing in the correct direction.
2. Arrange the arrows such that they follow on from one another without any gaps. The order of the arrows does not matter.
3. When you have finished, a single arrow that joins the starting and end points of your system will represent the resultant of the system of vectors.
4. Measure the length of this single arrow to find the magnitude of the resultant, and measure the angle from the horizontal or vertical to give its direction.

Perpendicular vectors

Key point: Vectors acting at 90° to one another are **independent**.

Thus, if several vectors are acting along each of two perpendicular lines of action, then each set of parallel vectors can be treated independently of one another and added as shown above. Once the situation has been simplified into two vectors acting perpendicular to each other then the resultant force can be found by either of the following methods:

- drawing a scale diagram of the vector triangle
- using Pythagoras's theorem and trigonometry

Wherever possible, the latter method should be used as it gives a more accurate answer.

Resolving vectors

Any vector can be resolved, using trigonometry, into two components acting perpendicular to one another; for example, a force F can be resolved into a horizontal component F_H and a vertical component F_V. This is useful for answering questions in which many vectors are acting in different directions. The two sets of perpendicular vectors can then be added and subtracted independently of one another.

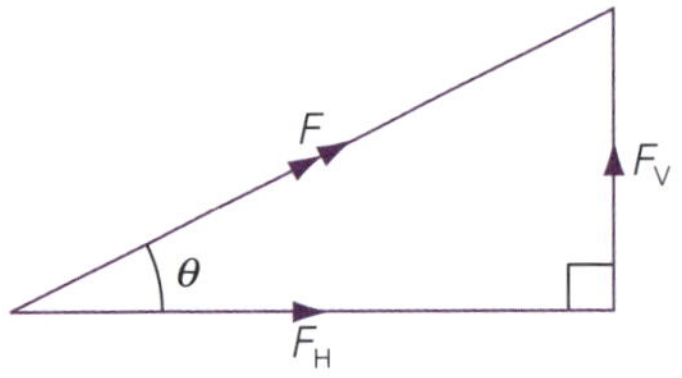

$F_H = F\cos\theta$

$F_V = F\sin\theta$

or

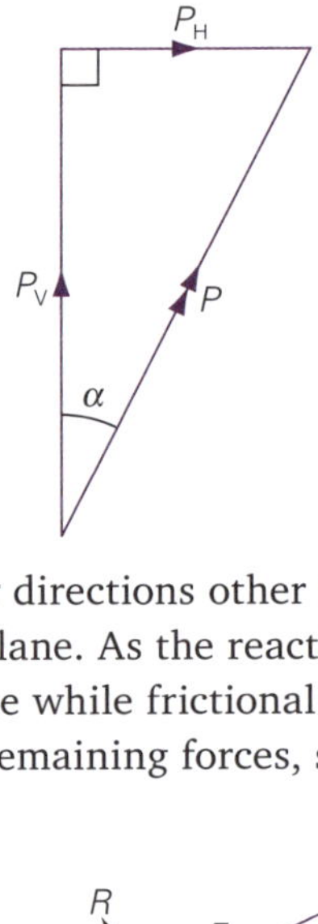

$P_H = P\sin\alpha$

$P_V = P\cos\alpha$

In some cases it is useful to resolve in a set of perpendicular directions other than horizontal and vertical, for example when an object is on an inclined plane. As the reaction between the object and the surface always acts perpendicular to the plane while frictional forces always act parallel to the plane, it is usually easiest to resolve any remaining forces, such as the object's weight, in these two directions.

$W_{parallel} = W\sin\theta$

$W_{perpendicular} = W\cos\theta$

Practice questions

1 What is the difference between a scalar and a vector quantity?

....

....

2 Complete the table below. One row has been done for you.

Quantity	SI unit	Scalar or vector?
Acceleration		
Density		
Tension		
Time	s	Scalar
Volume		

3 Over the course of a night, a slug travels in a complete circle, ending back at its starting point. If the radius of the circle is 3 m, find:

a its total distance travelled: ....

b its total displacement: ....

4 Find the magnitude and the direction of the resultant of the following pairs of vectors by drawing vector triangles:

a velocities of $24\,m\,s^{-1}$ vertically upwards and $10\,m\,s^{-1}$ horizontally to the right

b displacements of 4 km due north and 7 km due west

5 Check your answers to question 4 by calculations.

6 Find the magnitude and the direction of the resultant of the following displacements by drawing a vector triangle:

10 km due east and 5 km at a bearing of 307°

Magnitude ....

Direction ....

7 Resolve each of the following forces into horizontal and vertical components:

a 10 N at an angle of 30° above the horizontal

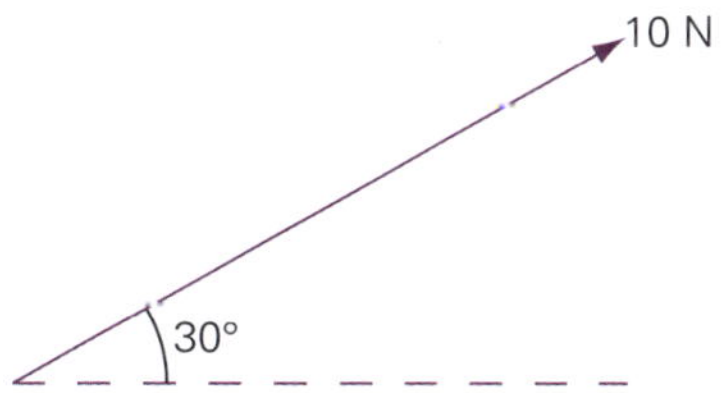

Horizontal ....

....

Vertical ....

....

b 35 N at an angle of 20° to the left of the vertical

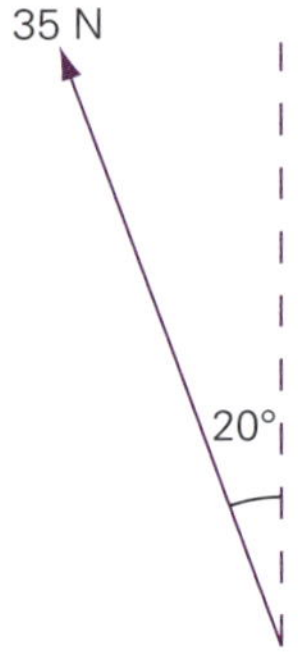

Horizontal ....

....

Vertical ....

....

8 An object on a slope has a mass of 5 kg. Taking $g = 9.81\,\mathrm{m\,s^{-2}}$:

a What is the object's weight?

....

b Resolve the object's weight into its components parallel and perpendicular to the slope in order to complete the diagram.

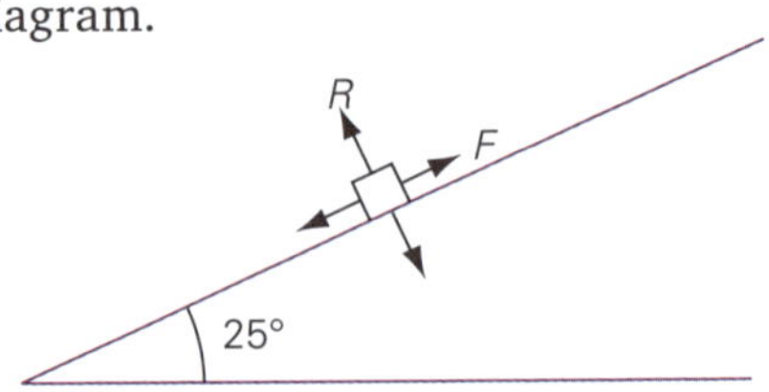

Parallel to slope ....

....

Perpendicular to slope ....

....

9 Decide whether each of the following statements is true or false. Beneath each of the correct statements write 'True'. For each of the incorrect statements, cross through the incorrect word or phrase and write the correct version in the space underneath.

a The resultant of two perpendicular vectors is always smaller in magnitude than each of the components.

....

b Two vectors acting at right angles to each other act independently.

....

c Mass is a vector and can therefore be added using normal arithmetic.

....

d Two vectors acting in opposite directions will always cancel each other out exactly.

....

e Acceleration is a vector; a negative acceleration indicates a deceleration.

....

f Two perpendicular forces of magnitude 9 N and 12 N combine to make a vector of magnitude 21 N.

....

g When an object is on a slope, it can be useful to resolve its weight into components parallel to the slope and vertically upwards.

....

10 Find the magnitude and direction of the resultant force acting on the object in this diagram.

(Hint: find the sum of the horizontal forces and the sum of the vertical forces separately, and then find the resultant of these.)

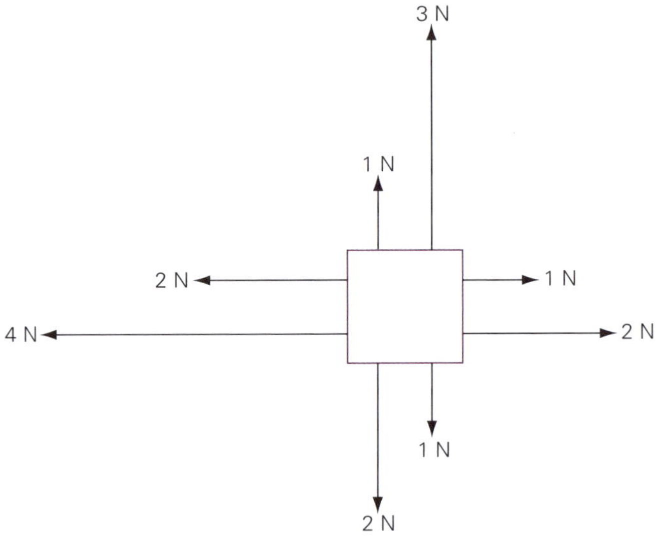

....

....

....

....

....

....

....

....

11 A particle with a weight of 49 N is suspended from two strings as shown in the diagram.

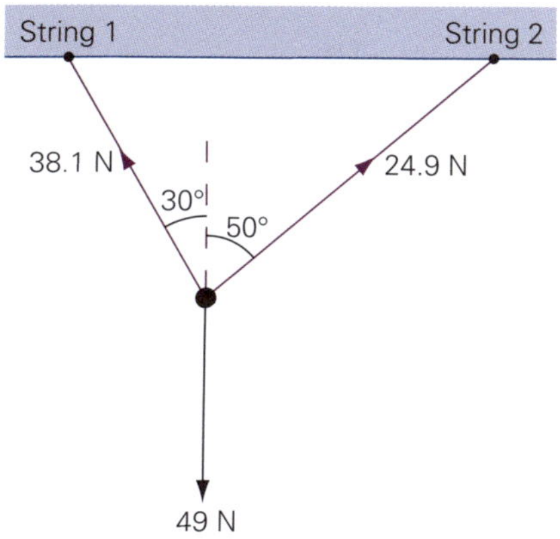

a Find the vertical component of:

i the tension in string 1

ii the tension in string 2

....

....

....

....

....

....

b Add the two values you obtained in part **a**. What do you notice?

....

....

c Find the horizontal component of

i the tension in string 1

ii the tension in string 2

....

....

....

....

....

....

d What do you notice about your two values from part **c**?

....

e This particle is in **equilibrium**. What do you think is one of the conditions required for equilibrium?

....

....

Topic 2 Equilibrium

Centre of gravity

Key point: The **centre of gravity** (sometimes called the centre of mass) of an object is the point where the entire weight of an object can be considered to act.

In a uniform, **regular** solid (e.g. a sphere or a cube), the centre of gravity is at the exact centre of the object. It is possible, however, to have an object where the centre of gravity is not within the object itself; for example, the centre of gravity of a banana may lie outside of the banana.

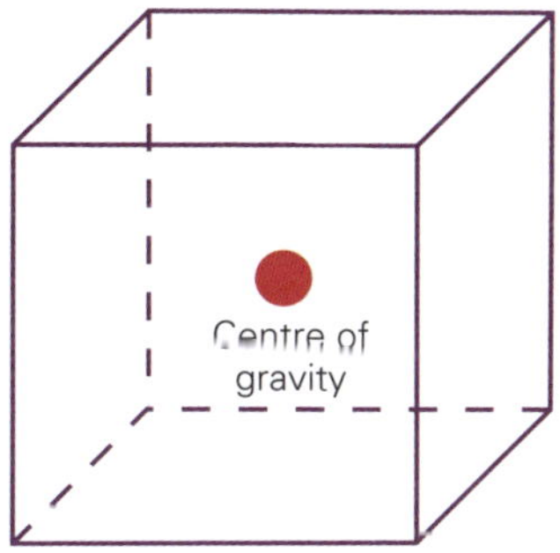

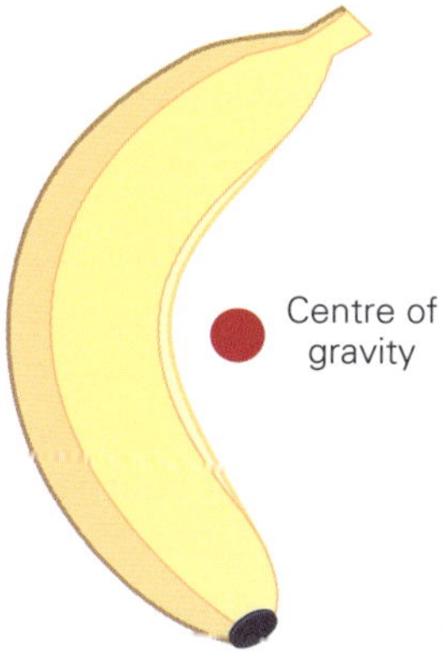

Finding the centre of gravity of an object

Key point: When an object is suspended so that it can turn freely, it hangs so that its centre of gravity is vertically below the point of suspension.

A plumb line can therefore be used to draw a line on which the centre of gravity must lie. By rotating the object, and repeating this procedure from a second suspension point (which must not be on the line you have just drawn), the exact position of the centre of gravity can be pinpointed — at the point of intersection of the two lines.

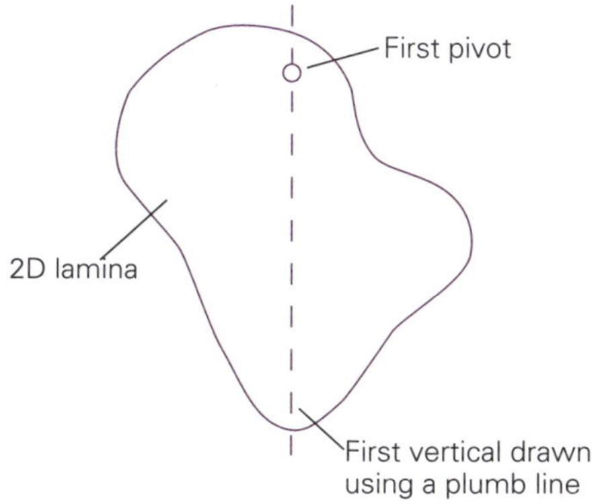

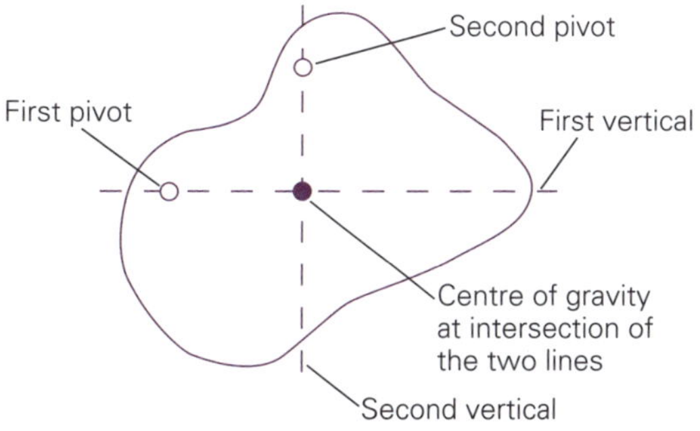

Torque, moments and couples

Torque

Consider an object that is free to rotate about an axis. When a force is applied that acts through a point not on the object's rotational axis, and in a direction other than parallel to the axis, it has a turning effect on the object. This turning effect is called a **torque**.

The torque created by a single force is called a **moment**, whereas a pair of anti-parallel forces of equal magnitude can produce a **couple**.

Key point: Moments and couples have units of N m. The direction of the turning effect must also be specified: either **clockwise** or **anticlockwise**.

Moments

When an object such as a rod or a ladder is on a **pivot** (represented by a triangle or an 'O') so that it can rotate freely, applying a force of magnitude F a **perpendicular distance** d from the pivot creates a turning effect called a **moment**. The size of the moment is given by:

$$\text{moment} = Fd$$

The resultant of a system of moments can be found much in the same way as finding the resultant force on an object. Find the sum of the clockwise moments, then the sum of the anticlockwise moments; the difference between these gives the resultant moment. Don't forget to state the direction.

Key point: A force has **no turning effect** if it does either of the following:

- passes through the axis of rotation or pivot — this can be useful when solving problems, as it means that we can eliminate an unknown force from the calculations by choosing to take moments about the point where it acts
- acts parallel to the axis of rotation

Non-perpendicular forces

Note that d in the moment formula is the **perpendicular distance to the line of action of the force**. For situations involving non-perpendicular forces, simply find the component of the force that *does* act at right angles to the rod. For AS physics, it is unlikely that you will be asked to do this, although you must always explain the full meaning of d if asked to define torque, moments or couples.

Couples

When two forces that are equal in magnitude but anti-parallel (have opposite directions) act on an object, they are referred to as a **couple**. Since the forces cancel each other out in every direction, they cannot generate a resultant force and so produce **rotation only**, rather than a translational effect. The size of the couple is given by:

couple = Fd

where F is the magnitude of each of the forces and d is the perpendicular distance between their lines of action, independent of the position of the rotational axis.

Conditions for equilibrium

Key point: There are two conditions required for an object to be in equilibrium:

1 the **resultant force** acting on the object **is zero in all directions**

2 the **resultant moment** acting on the object **about any axis is zero**

It may be necessary to use both of these conditions in order to solve problems involving systems in equilibrium.

Closed triangles of forces

Condition number one states that for a system to be in equilibrium, the resultant of the system of forces must be zero. An unbalanced system (one that is not in equilibrium) that has a resultant force R can be balanced by introducing a further force equal in magnitude to R but which acts in the opposite direction.

Recall the vector triangles considered on page 3. Questions that ask for the force required to establish equilibrium can be solved using a modified version of this method, which is known as creating a **closed triangle of forces**. To do this, draw a scale diagram of the forces such that they follow on from one another. The force required to keep the particle in equilibrium is given by the line needed to close the triangle. Note the direction of this force: it is *opposite* to that of the resultant of the two original forces.

The principle of moments

Condition number two for equilibrium is a version of the **principle of moments:**

For a body to be in equilibrium, the sum of the clockwise moments about any axis must be equal to the sum of the anticlockwise moments.

Practice questions

1 Define the following terms:

Hint: Often the best way to define a physics term is to state the equation that governs it and then define all of the terms in the equation, giving their units. Drawing a labelled diagram will often gain partial marks in an examination or, at least, make it easier for an examiner to understand what you have written.

a centre of gravity

....

....

b the moment of a force

....

....

....

c couple

....

....

....

d principle of moments

....

....

2 Would you classify the moment of a force as a scalar or a vector quantity? Explain your answer.

....

....

3 Find the magnitude and direction of the moment about the point O created by the forces in each of the following diagrams.

a

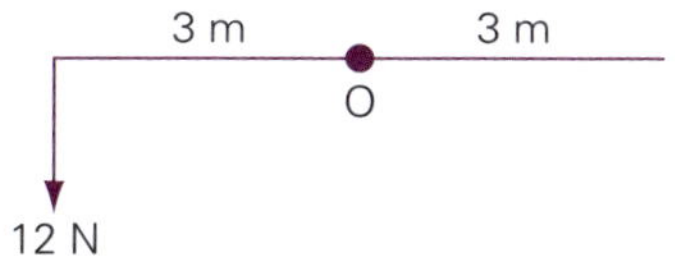

....

....

b

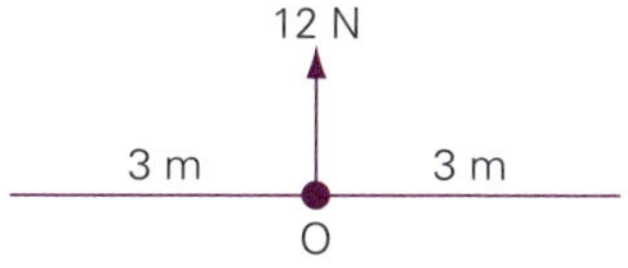

....

....

4 What is the magnitude and direction of the net moment about the point O in the following diagram, *if it is drawn to a scale of 1:100*?

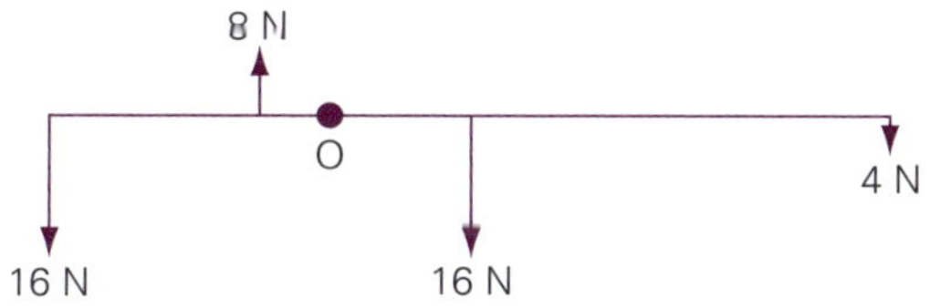

....

....

....

....

5 What is the magnitude and direction of the moment about the pivot created by the force shown in this diagram?

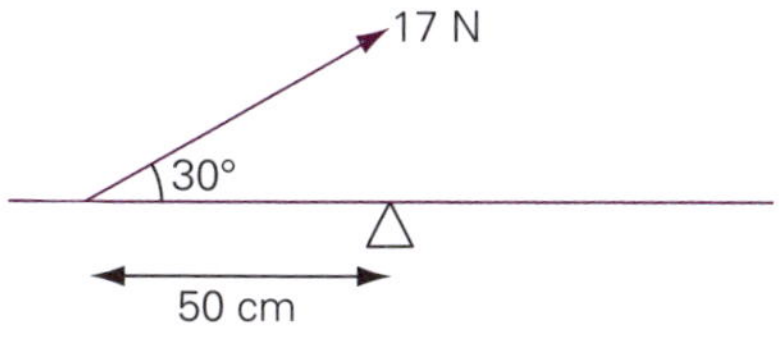

....

....

....

....

6 Find the size and direction of the couples in the following two diagrams.

a

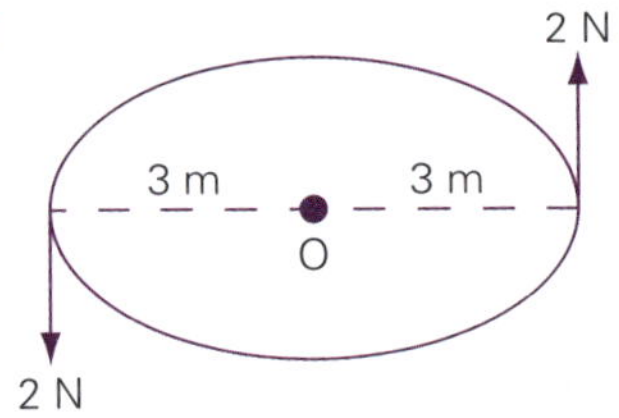

b

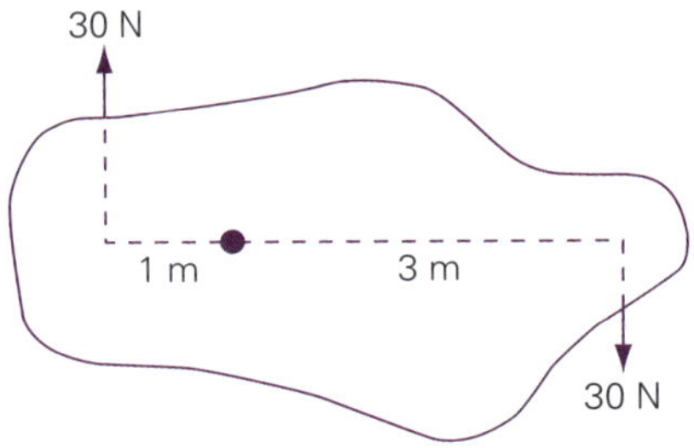

7 Draw a labelled diagram to show the magnitude and direction of a pair of forces a perpendicular distance of 4 cm apart that create a clockwise couple of 0.36 N m.

8 What are the two conditions required for equilibrium?

9 A particle of mass 353 g is held in equilibrium by a force of magnitude 2 N acting horizontally to the right and a force P that acts at an angle θ to the left of the vertical.

a Find the magnitude of P and the value of θ by drawing a closed vector triangle to scale.

b Check your answers to part **a** by resolving the forces in two perpendicular directions.

....

....

....

....

....

10 The system in the diagram below is in equilibrium. By taking moments about a suitable point, find the magnitude of *R*.

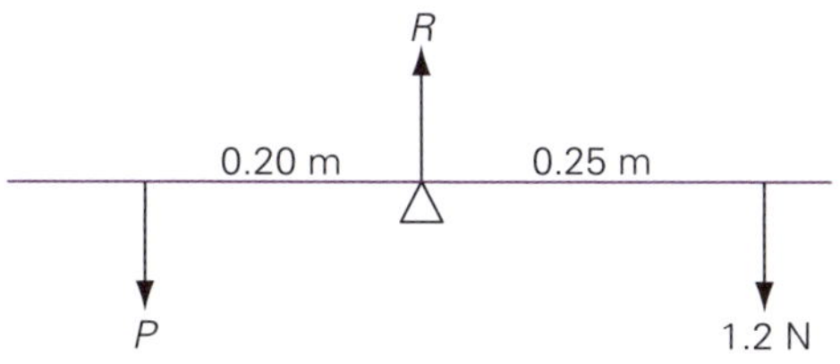

....

....

....

11 A painter carries a uniform, regular plank that is 2 m long and has a mass M balanced on his shoulder. The painter's shoulder exerts an upwards reaction force R on the plank a distance of 0.85 m from one end, and his hand exerts a downward force F a distance of 0.4 m in front of his shoulder.

a What is acting as a pivot?

....

b Draw a free-body diagram to show what is happening.

c If the system is in equilibrium, show that:

i $R = Mg + F$

....

ii $F = \dfrac{3Mg}{8}$

....

....

....

d If $M = 5\,\text{kg}$, find the magnitude of:

i F

....

....

ii R

....

Topic 3 Kinematics

Important definitions and equations

Kinematics is the study of situations involving movement. You need to be able to recall and use the following definitions and equations:

Distance travelled is a scalar quantity that tells you the total distance covered by an object during a journey.

Displacement is a vector indicating the length and direction of the shortest path from one point to another.

Key point: Displacement is denoted by *s*, rather than the *d* that you may have used earlier in your studies; *d* may still be used for the scalar equivalent of displacement — the distance between two points — but kinematics questions will always use *s*.

Speed (a scalar) is the rate of change of an object's position.

Velocity (a vector) is the rate of change of an object's displacement:

$$v = \frac{\Delta s}{t}$$

Acceleration (a vector) is the rate of change of an object's velocity:

$$a = \frac{\Delta v}{t}$$

Notation: The symbol Δ ('delta') can be read as 'change in'.

Note: Some specifications use Δt rather than t alone in the above formulae; however, the position of time $t = 0$ can be defined such that the two quantities are equivalent.

Instantaneous and average values

If a car takes 2 hours to travel a total distance of 140 km, we can say that its **average speed** was $70\,\text{km}\,\text{h}^{-1}$. However, this does not mean that the car travelled at $70\,\text{km}\,\text{h}^{-1}$ for the whole journey: it may have been travelling faster at some points and slower at others; that is to say, we cannot tell what its **instantaneous speed** would have been at any given time.

Notation: A horizontal bar over the top of a letter indicates an average value, e.g. $\bar{v}$ represents average velocity.

Graphical representation

Displacement–time graphs

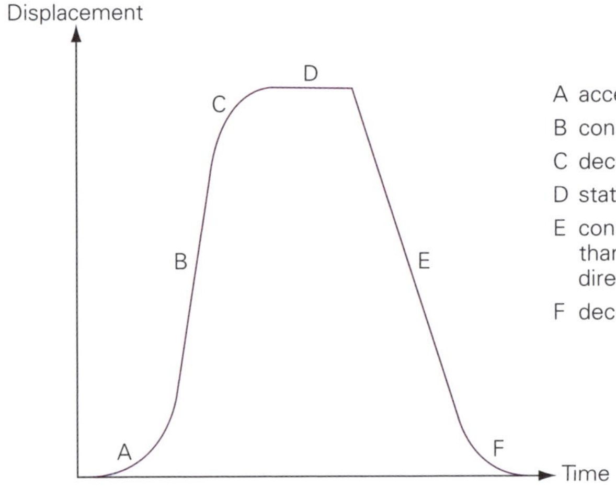

A acceleration
B constant velocity
C deceleration to rest
D stationary
E constant velocity, slower than and in opposite direction to B
F deceleration

Key points:

- The **gradient** of the tangent to the graph at any point gives the **instantaneous velocity** of the object; the steeper the tangent line, the faster the object is moving.
- A horizontal line means that the object is stationary.
- A constant gradient indicates constant velocity.
- A positive gradient indicates motion in the 'positive' direction; a negative gradient indicates motion in the 'negative' direction.
- An increasing gradient indicates acceleration; a reducing gradient indicates deceleration.
- A negative value for the displacement indicates a position in the negative direction.
- The **average speed** for the whole journey or over a given time period can be found using the formula:

$$\text{average speed} = \frac{\text{distance travelled}}{\text{time taken}}$$

Distance–time graphs

These are similar to displacement–time graphs with only a couple of differences.

Key differences:

- The *gradient* of the tangent to the graph at any point gives the *instantaneous speed* of the object.
- There can never be a negative gradient as the distance travelled by an object cannot decrease over time.

Velocity–time graphs

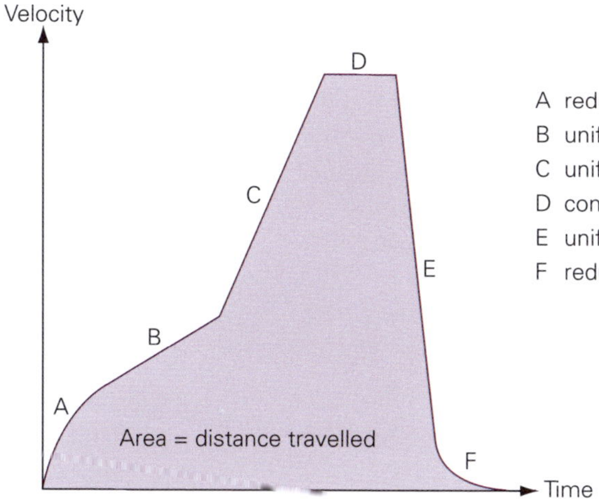

A reducing acceleration
B uniform acceleration
C uniform acceleration, greater than B
D constant velocity
E uniform deceleration, greater in magnitude than B or C
F reducing deceleration to rest

Key points:

- The **gradient** of the tangent to the graph at any point gives the **instantaneous acceleration** of the object.
- A horizontal line means that the object is travelling at a constant velocity.
- A constant gradient indicates uniform acceleration.
- A positive gradient indicates acceleration; a negative gradient indicates deceleration.
- An increasing gradient indicates increasing acceleration; a reducing gradient indicates reducing acceleration.
- The **distance travelled** is given by the **magnitude of the area enclosed** by the graph and the time axis for any given time interval.
- The **average speed** for the whole journey or over a given time period can then be found using the formula:

$$\text{average speed} = \frac{\text{distance travelled}}{\text{time taken}}$$

Negative velocities

A negative value for the velocity indicates movement in the negative direction — for example, an object moving backwards, or moving downwards.

In order to calculate the total **displacement** in a given time interval, you must take into account whether the motion is in the positive or the negative region of the graph:

total displacement = area in positive region – area in negative region

The four equations of motion

The four equations of motion are often referred to as *'suvat'* equations. These are a group of equations that can be used to solve almost any kinematics question involving uniform acceleration. You need to be able to derive and use each of them. Each equation contains four of the following five terms:

displacement	s
initial velocity	u
final velocity	v
acceleration	a
time (taken)	t

1 $v = u + at$ (s is not involved)

2 $s = \frac{(u+v)}{2}t$ (a is not involved)

3 $s = ut + \frac{1}{2}at^2$ (v is not involved)

4 $v^2 = u^2 + 2as$ (t is not involved)

Key point: Which equation you use will depend on the known and unknown quantities in the problem. For example, if you know v, u and s and want to find a, you need to use the equation that involves these four quantities but not t, which is Equation 4: $v^2 = u^2 + 2as$.

All these equations can be derived from first principles. For example, Equation 3 can be derived by considering the area beneath a velocity–time graph showing uniform acceleration. You could be asked to perform these derivations in an exam, so take time to familiarise yourself with them now.

The above four equations of motion deal with motion in one direction. Thus for many applications it is necessary to resolve the object's velocity, acceleration and displacement into two perpendicular directions, usually horizontal and vertical. In this book, the subscripts x and y are used to denote components of the motion in the horizontal and vertical directions, respectively; for example, v_x denotes final horizontal velocity.

Projectiles

A **projectile** is an object that is moving freely under the influence of gravity and/or air resistance. In circumstances where air resistance is considered to be negligible and the object is thus under the influence of gravity alone, the object is said to be in **free fall**, regardless of its direction of motion. For AS and A2 physics, this free fall assumption will always be made when modelling projectile motion.

Key point: By convention, **the upwards direction is taken to be positive**. Thus if a ball *falls* through a height of 5 m, we say $s = -5$ m. Similarly, because weight acts in the downwards direction, and all objects in free fall accelerate at the same rate, we take $a = -g$ for all projectiles questions.

When answering questions that involve projectiles, the following steps should be followed:

1. Draw a labelled diagram of the situation.
2. Resolve any known velocities and/or displacements into their components parallel and perpendicular to the ground.
3. **Apply suitable equations of motion in each direction independently.**
   - x components: there is no acceleration in the horizontal direction, so only Equation 2 is necessary, giving $s_x = u_x t$.
   - y components: in the vertical direction, choose the appropriate *suvat* equation, remembering to use $a = -g$.
4. If necessary, use the fact that t represents the same quantity in both directions.

Key points:

- The **maximum height** reached, h, depends only on the initial vertical velocity of the projectile. Equation 4, with $v = 0$ and $a = -g$, can be rearranged to give $u_y^2 = 2gh$.
- The **time of flight** can be found by solving Equation 3 with the appropriate value for s_y. Note that this is a quadratic equation and it will therefore yield two values of t — a little thought should tell you which is the correct one.
- To find the total **horizontal distance travelled**, the time of flight must be found as above, and then substituted into the equation for horizontal motion.

Practice questions

1 A cheetah can reach top speeds of up to 108 km h^{-1}.

a What is this speed in m s^{-1}?

....

....

b If the cheetah runs at this speed for 20 s, how far does it travel?

....

c Explain why the actual distance travelled by a cheetah over 20 s would be less than the value you calculated in part **b**.

....

....

d If the cheetah accelerates from rest to 72 $km\,h^{-1}$ in 2 s, what is its acceleration in $m\,s^{-2}$?

....

....

....

2 Fred runs a 400 m race in 50 s, completing one full lap of a 400 m oval track.

a At the end of the race, what was Fred's

i distance travelled?

....

ii displacement from his starting position?

....

b What was Fred's average speed?

....

c What was Fred's instantaneous speed after 20 s? Choose from the following options and explain your choice.

A Less than his average speed.

B Equal to his average speed.

C More than his average speed.

D We cannot say — we would need more information.

....

....

3 Complete this table explaining the features of displacement–time graphs and velocity–time graphs.

Feature of graph	Displacement–time graphs	Velocity–time graphs
Finding the gradient gives the...		Instantaneous acceleration
A constant gradient means...		
A horizontal line means...	The object is stationary	
A negative gradient means...		
The area under the graph represents the...	N/A	

4 Sketch a displacement–time graph to show each of the following on the axes provided:

a a stationary object

b constant non-zero velocity

c deceleration

a

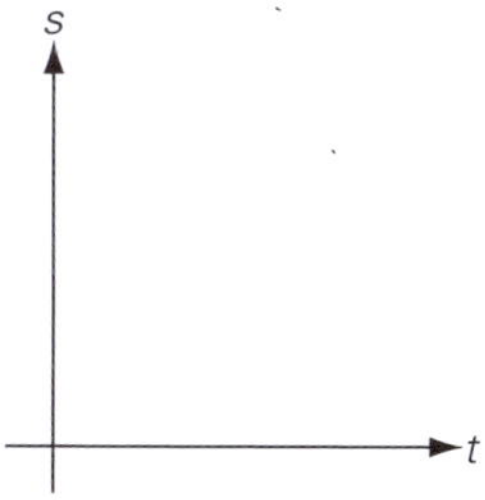

b

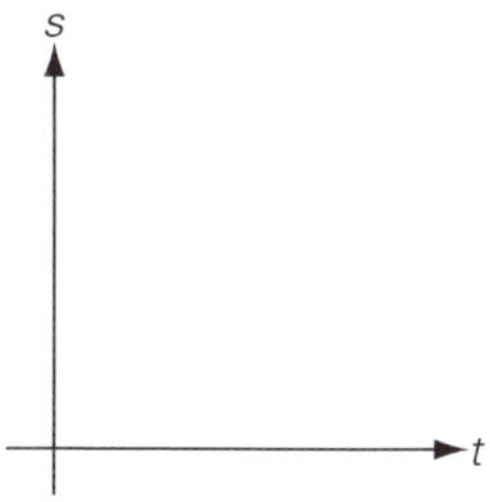

c

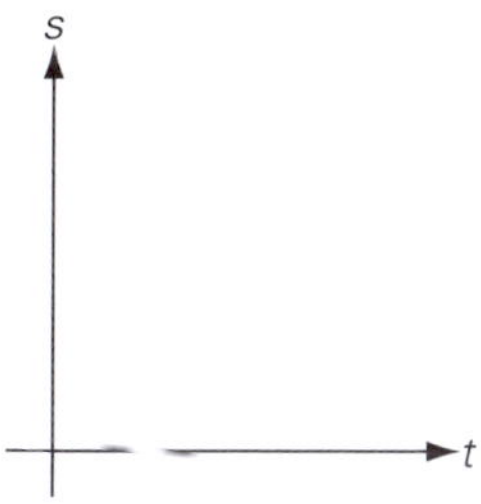

5 Sketch a velocity–time graph to show each of the following on the axes provided:

a a stationary object

b uniform non-zero acceleration

c increasing acceleration

a

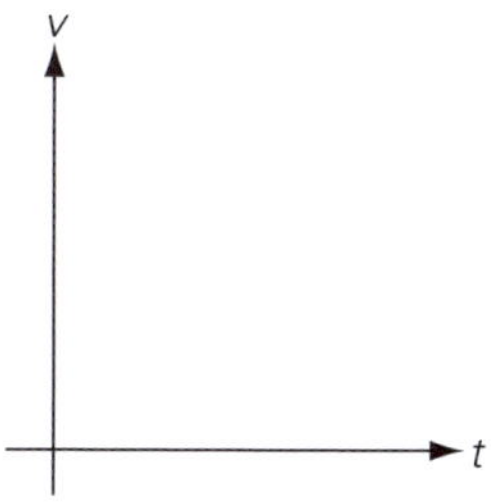

b

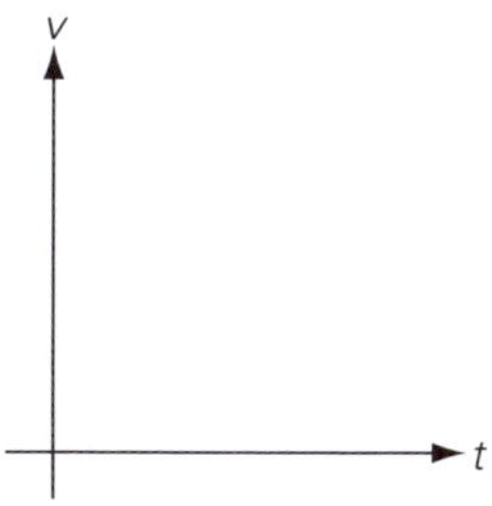

c

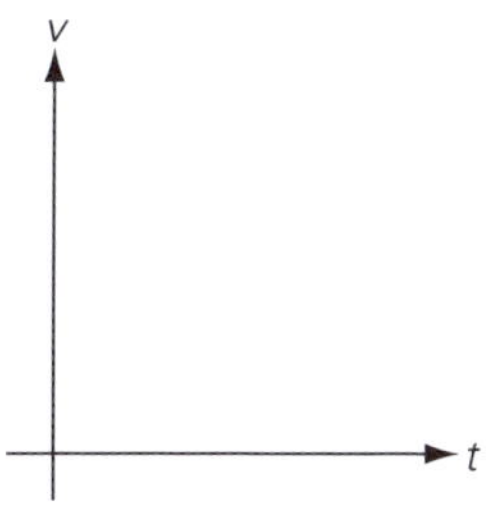

6

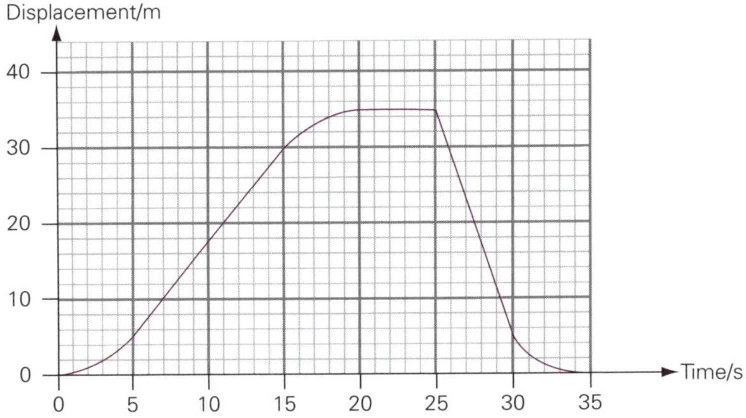

a Between which set(s) of times is the object in the diagram:

i accelerating?

....

ii decelerating?

....

iii travelling at a constant non-zero velocity?

....

iv stationary?

....

b What is the maximum velocity reached by the object?

....

....

....

c What is the object's maximum displacement from its starting position?

....

....

7 Use the correct equation of motion to answer each of the following questions.

a Find u if $v = 34\,\text{m}\,\text{s}^{-1}$, $a = 3.5\,\text{m}\,\text{s}^{-2}$, $t = 4\,\text{s}$.

b Find s if $u = 10\,\text{m}\,\text{s}^{-1}$, $v = 20\,\text{m}\,\text{s}^{-1}$, $a = 9.8\,\text{m}\,\text{s}^{-2}$.

c Find a if $s = 294.6\,\text{m}$, $u = 41\,\text{m}\,\text{s}^{-1}$, $t = 6\,\text{s}$.

d Find t if $s = -23.4\,\text{m}$, $u = 5\,\text{m}\,\text{s}^{-1}$, $v = -18\,\text{m}\,\text{s}^{-1}$.

8 A car accelerates uniformly from rest at $0.25\,\text{m}\,\text{s}^{-2}$ for 60 s, maintains a constant velocity for 40 s and then accelerates for a further 10 s to reach $25\,\text{m}\,\text{s}^{-1}$. The car maintains this top speed for 30 s before decelerating to rest over a further 40 s.

a What is the total time taken for this journey?

b On the axes given, plot a velocity–time graph for this journey.

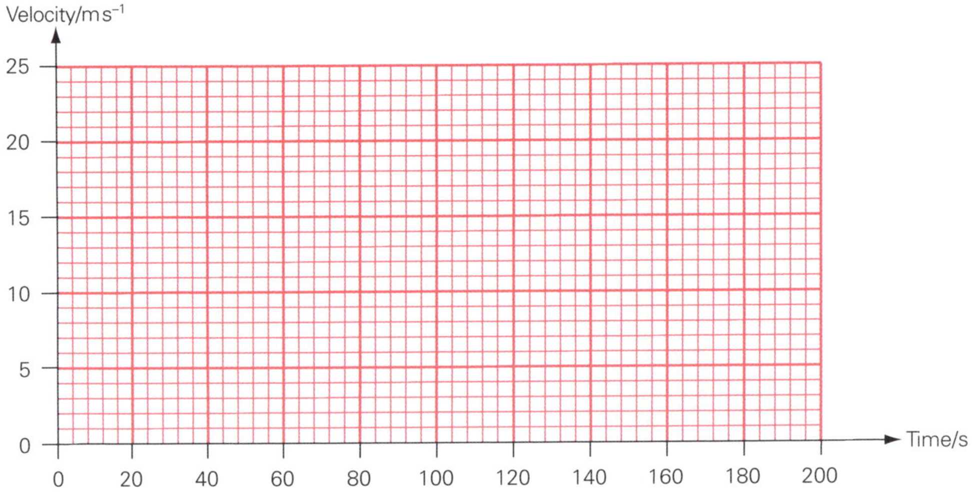

c What is the car's instantaneous velocity after 160 s?

....

d What is the car's instantaneous acceleration after 160 s?

....

....

e What is the distance travelled by the car

i during the first 60 s?

....

ii between 60 s and 100 s?

....

iii between 100 s and 110 s?

....

f What was the average speed of the car during the first 110 s?

....

....

9 A low-flying aircraft, flying horizontally with a speed of $100\,m\,s^{-1}$ at a height of 200 m above the ground, drops an aid parcel. Assuming that air resistance is negligible, find:

a the time taken for the parcel to hit the ground

....

....

....

b the horizontal distance travelled by the parcel as it falls

....

....

c the speed of the parcel just before it hits the ground

....

....

....

....

d the direction of the parcel's velocity just before hitting the ground

....

....

....

10 A cannonball is fired at an angle of 30° to the horizontal with a speed of $40\,m\,s^{-1}$. Assuming that air resistance is negligible, find:

a the horizontal and vertical components of the cannonball's initial velocity

....

....

b the maximum height reached by the cannonball

....

....

....

....

c the length of time for which the cannonball is in the air

....

....

....

....

d the horizontal distance travelled by the cannonball before it hits the ground

....

....

....

11 The velocity–time graph below is for an object thrown vertically upwards on the moon.

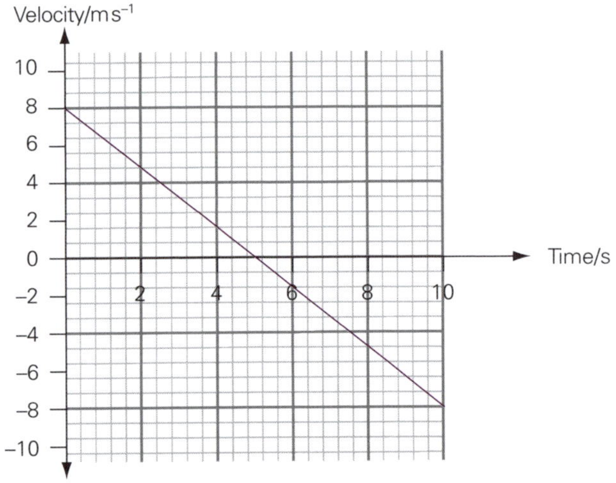

a State the magnitude and direction of the object's initial velocity.

....

b Describe the object's motion:

i between 0 and 5 s

....

ii between 5 and 10 s

....

c After how long does the object reach its maximum height above the ground?

....

d Using the graph, calculate the maximum height reached by the object.

....

....

....

e Explain, with the aid of a calculation, how you can tell that this scenario did not take place on Earth.

....

....

....

f What is the total displacement of the object from its starting position after 7.5 s?

....

....

....

....

Topic 4 Forces

Newton's laws

Newton's first law

A body stays at rest or, if it is moving, continues at a constant velocity in a straight line *unless* it is acted upon by an external resultant force.

Summary: If no external resultant force acts on the body:

- a stationary body remains at rest
- a body moving with a velocity v continues at the same speed and in the same direction

Newton's second law

The rate of change of momentum of an object is directly proportional to the resultant force acting upon it.

This law describes what happens if there *is* a resultant force acting on a body; it is commonly summarised as the well-known equation $F = ma$.

Note: At very high, 'relativistic', speeds — close to the speed of light — the mass of an object is no longer constant; it increases. Therefore Newton's second law becomes a bit more complicated, and $F = ma$ no longer applies. You should be aware of this limitation, although you will not be asked to use Newton's second law in relativistic situations for AS physics.

Summary:

- An external resultant force acting on a body causes the body to **accelerate in the direction of the force**.
- In situations where mass is constant, Newton's second law becomes $F = ma$, where F = **resultant** force acting on the body, m = mass of body, and a = consequential acceleration.
- The newton is the force required to cause a body of mass 1 kg to accelerate at $1\,\text{m}\,\text{s}^{-2}$. Thus $\mathbf{1\,N = 1\,kg\,m\,s^{-2}}$.

Newton's third law

If a body A exerts a force *F* on a second body B, then B exerts a force on A which is equal in magnitude to *F* but acts in the opposite direction.

Key point: The forces of each action–reaction pair **act on different bodies**, and so do not cancel each other out.

Summary:

- Forces come in anti-parallel (parallel but acting in opposite directions) pairs of equal magnitude.
- When two bodies interact, one of the pair of forces acts on each of the bodies.

Terminal velocity

Objects accelerating through fluids experience a drag force that increases as their velocity increases. As a result, their acceleration decreases; they reach terminal velocity when the drag force is exactly equal and opposite to the force causing their initial acceleration. Common examples of terminal velocity being approached are skydivers — both before and after opening their parachutes — and cars reaching their maximum speeds.

You need to be able to recognise and sketch displacement–time and velocity–time graphs showing terminal velocity being reached.

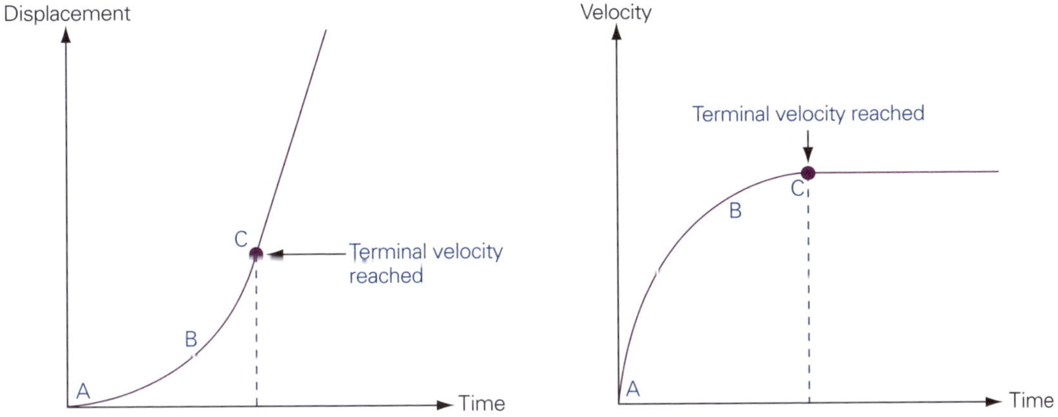

You also need to be able to describe this motion with reference to the forces that are acting. The key stages are as follows:

A Drag is zero or very small ⇒ large resultant force ⇒ high acceleration.

B Resisting force increases ⇒ resultant force is decreasing → acceleration is decreasing.

C Resisting force is equal in magnitude to the forward force ⇒ no resultant force ⇒ no further acceleration so terminal velocity is reached.

Key point: A common student error is to say that 'the *velocity* is reducing until it reaches a steady value'. This is, of course, nonsense: it is the **rate of change of velocity** that is reducing, not the velocity itself. Therefore, when asked to describe this motion, you should say that 'there is initially a high acceleration, but this **acceleration reduces gradually to zero**, at which point a **constant — terminal — velocity is reached**'.

Practice questions

1 In your own words, write a short paragraph to explain the meaning of each of Newton's laws of motion. Include diagrams where they would be helpful.

....

....

2 Define the following terms:

a newton (N)

b *g*

c weight

3 In all of the following diagrams, the body is *moving up the page*. For each of the forces shown, describe its effect — if any — on the body's speed and direction.

a

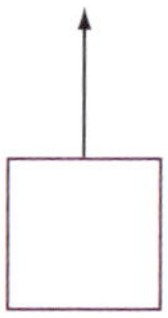

b

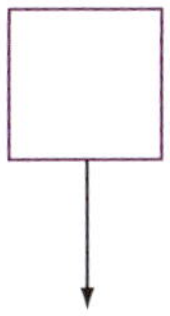

c

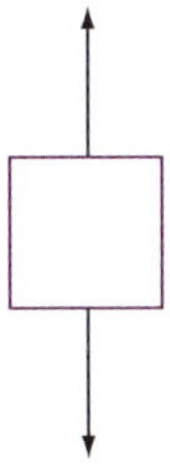

d

4 An object is put on a frictionless surface which is inclined at an angle θ to the horizontal. What is the value of θ if the object accelerates down the slope at a rate of $6.31\,\text{m}\,\text{s}^{-2}$?

5 A train with a mass of 1000 tonnes is travelling at $90\,\text{km}\,\text{h}^{-1}$. What frictional force must be provided by its brakes if it is to stop in 40 seconds? (1 tonne = 1000 kg)

6 The object in the diagram is in horizontal equilibrium and has a weight of 98.1 N.

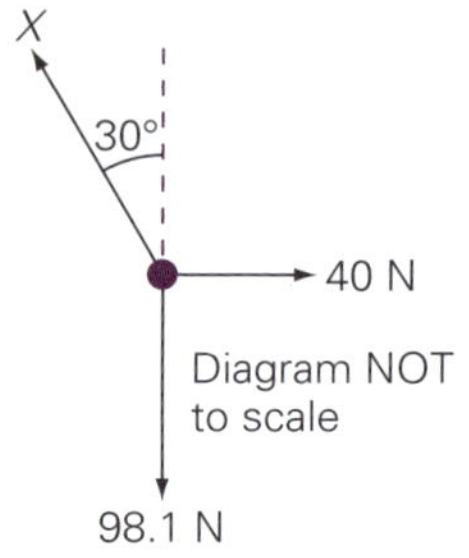

a Find the magnitude of the force X required to maintain horizontal equilibrium.

....

....

b What is the magnitude and direction of the object's acceleration in the vertical direction?

....

....

....

....

....

7 A car engine provides a constant forward thrust of 2 kN. The car's mass is 2.5 tonnes.

a What is the maximum acceleration that the engine could provide?

....

b State two factors that will affect the size of the air resistance experienced by the car when it is moving.

....

c What will be the car's acceleration when the resistive forces reach 1.2 kN?

....

....

8 Fill in the missing words in this paragraph describing the motion of a skydiver:

a When a skydiver jumps out of a plane, the only force acting on him in the vertical direction is his ...., which acts vertically downwards towards the .... of the Earth. There is therefore a large .... causing him to .... downwards; he accelerates initially at .... $m\,s^{-2}$.

b As he falls, his .... increases and the magnitude of the .... he experiences increases as a result. The resultant force therefore ...., as does the skydiver's ....

c Eventually, the air resistance becomes equal to the skydiver's ...., and he reaches a ....

d At a suitable height, the skydiver opens his parachute. There is now a resultant force in the .... direction as the parachute has a large ...., increasing the air resistance.

e This resultant force causes the skydiver to ....; as he does so, the magnitude of the .... decreases.

f The size of the resultant force decreases until it is once again equal to the skydiver's ...., whereupon he reaches a second .... that is safe to land at.

9 Sketch a velocity–time graph for the skydiver in Question 8, and label the points/stages **a** to **f**.

Topic 5 Energy

Potential energy

Potential energy is **gained by doing work against a force**. For example, a person climbing a ladder is doing work against his/her weight — the force of gravitational attraction due to the Earth — and is therefore gaining gravitational potential energy (GPE). By the relation:

work done = energy transferred

and assuming that the system is 100% efficient, the amount of potential energy *gained* is equal to the work done *against* the force.

Conversely, when a force does work *on* an object, the object *loses* potential energy as it is converted into other forms. For example, when a child goes down a slide, gravitational potential energy is converted predominantly into kinetic energy, with some energy being dissipated as thermal energy via frictional forces.

Work done

The amount of work done *by* or *against* a force is given by:

work done by a force = magnitude of the force × displacement parallel to the force
work done against a force = magnitude of the force × displacement anti-parallel to the force

The joule (J) is defined as the work done on an object when moving it a distance of 1 m against a force of 1 N. Thus **1 J = 1 N m**. The joule is the unit of all types of energy in addition to work done.

When a mass m is raised through a height h close to the surface of the Earth, work is done against the mass's weight $W = mg$; the mass therefore gains gravitational potential energy (GPE). This is the derivation of the formula for GPE that you may already know:

change in GPE = weight × change in height

that is, $\Delta E_{grav} = mg\Delta h$

Note that this relation is only valid close to the Earth's surface where g can be taken as a constant value. You will learn about gravitational potential energy in a more general context at A2.

Non-parallel forces

When the force and the displacement are at an angle θ to one another, the component of the displacement along the line of action of the force must be found.

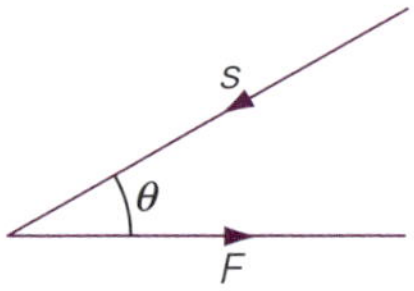

The component of s along the line of action of F is given by $s\cos\theta$. Thus the equation for work done can be fully expressed as:

work done $= Fs\cos\theta$

where θ is the angle between F and s.

Kinetic energy

All moving objects possess kinetic energy. The amount of kinetic energy is given by:

$$E_k = \frac{1}{2}mv^2$$

where m is the mass of the object and v is the magnitude of the object's velocity (i.e. the object's speed).

Conservation of energy

The law of conservation of energy states that the **total amount of energy of a closed system remains constant.**

Key points:

1. The energy is allowed to change forms (e.g. kinetic energy → GPE).
2. The energy is allowed to be transferred between objects within the system.
3. No energy is allowed to enter or exit the system.

Collisions

In an **elastic** collision, there is no loss of kinetic energy. The conservation of energy statement then becomes:

total kinetic energy before collision = total kinetic energy after collision

In real (inelastic) collisions, some energy is always transferred as thermal energy to the bodies involved and to the surroundings, dissipated as sound energy, or converted to potential energy if there is permanent deformation of the object(s) in the collision. Accounting for this, however, the total amount of energy before and after the collision is always the same.

Falling and rising

It is often simplest to answer free fall questions by considering the conservation of energy.

If E_{grav} is taken to be zero at ground level, then energy considerations can be used to find the change in velocity of an object when it falls through a height Δh:

E_k gained $= E_{grav}$ lost

$\therefore \frac{1}{2}m(\Delta v)^2 = mg\Delta h$

Hence $\Delta v = \sqrt{2g\Delta h}$

The equation $\frac{1}{2}m(\Delta v)^2 = mg\Delta h$ can also be used to find the maximum height reached by a projectile with initial vertical velocity u:

$$h_{\max} = \frac{u^2}{2g}$$

It should be noted that this is the same result as could be obtained from the equation of motion $v^2 = u^2 + 2as$ with $v = 0$ and $a = -g$.

Key point: The mass cancels from both sides of the equation for free fall energy equivalence; thus the mass does not need to be known in order to solve such problems.

Power and efficiency

Power

Power is the rate of doing work:

$$\text{power} = \frac{\text{work done}}{\text{time taken}} = \frac{\text{energy transferred}}{\text{time taken}} \quad \text{i.e. } P = \frac{\Delta E}{t}$$

The watt (W) is a rate of energy transfer equivalent to 1 joule being transferred per second:

$1\,\text{W} = 1\,\text{J}\,\text{s}^{-1}$

Power against a force

It can be shown that the power P required to maintain an object's velocity v against a force F is given by $P = Fv$.

Efficiency

Efficiency is the ratio of *useful* energy output of a system to the total energy input; it is usually expressed as a percentage:

$$\text{efficiency} = \frac{\text{useful energy out}}{\text{total energy in}} \times 100\%$$

The ratio of useful power output to total power input can also be used.

Key point: No system is truly 100% efficient because some energy is always *dissipated*, usually as thermal energy.

Practice questions

Unless specified otherwise, assume that air resistance is negligible for all the questions in this section.

1 There is something wrong in each of the following statements. For each one, cross through the incorrect word and write the correct word or phrase in the space underneath.

a Work has to be done on an electron to bring it closer to a positively charged plate.

....

b When two similarly charged particles are brought closer together, they both gain gravitational potential energy.

....

c Work has to be done against an object's mass to raise it from ground level to a height of 100 m.

....

d 1 J is the energy required to lift a mass of 1 kg through a height of 1 m close to the surface of the Earth.

....

e It takes more energy to lift a mass through a given height if you do it faster.

....

f If an object is dropped from twice the height then it will have twice the velocity when it hits the ground.

....

2 Define the following terms:

a closed system

....

....

b dissipate

....

....

c efficiency

....

....

d 1 watt

....

....

3 Calculate the following, paying attention to the units given:

a the kinetic energy of a 10 g mouse running at 0.5 m s^{-1}

....

....

b the speed of a car of mass 1800 kg which has a kinetic energy of 32.4 kJ

....

....

....

c the increase in gravitational potential energy of a 50 kg girl when she climbs a height of 40 m

....

....

d the height through which a 20 kg mass must be raised in order to gain 4.5 kJ of gravitational potential energy

....

....

e the work done against a constant frictional force of 26 N over a distance of 120 m

....

....

f the work done against gravity when walking a horizontal distance of 73 m

....

g the force required to bring a 5 kg object with a velocity of 2 m s^{-1} to rest over a distance of 8 cm

....

....

....

h the work done when moving a distance of 2 m against a force of 6 N which acts at an angle of 20° to the direction of motion

....

....

....

i the angle between a force of magnitude 0.63 N and a body's displacement if moving the body through a distance of 12 cm requires 0.068 J of work to be done

....

....

....

4 A mass m falls from a height h to ground level. On a single set of axes, sketch two lines to show how the gravitational potential energy and kinetic energy of the mass each vary with the height above the ground.

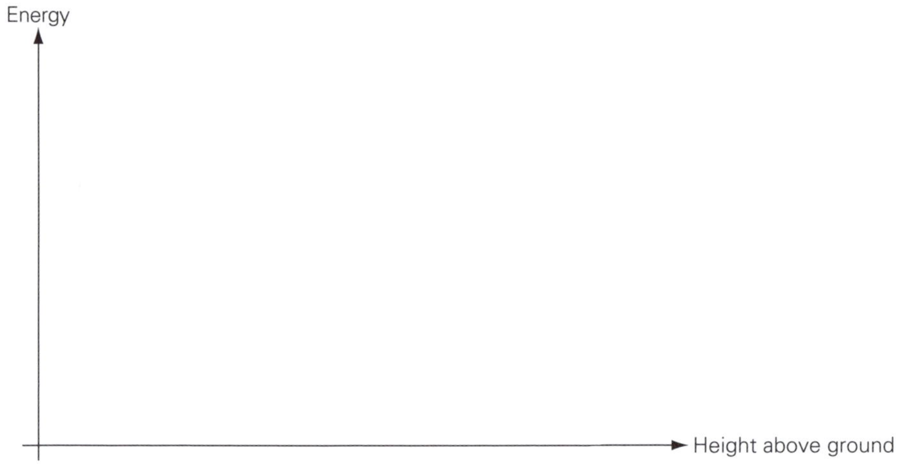

5 A pebble is dropped from a height of 80 m. What will be its velocity:

a just before it hits the ground?

....

....

....

b after falling 40 m?

....

....

....

c 10 m above the ground?

....

....

....

6 A football is kicked at an angle of 40° to the horizontal. What initial velocity is required for the ball to reach a maximum height of 10 m above the ground?

7 A body of mass m starts from rest on a smooth slope inclined at 60° to the horizontal. What is its speed when it has moved a distance of 5 m down the slope?

8 By considering the formulae for work done and for power, show that the power P required to maintain an object's velocity v against a force F is given by $P = Fv$.

9 Calculate:

a the power required to lift 20 objects each weighing 5 kg through a height of 1.2 m in 48 s

....

....

....

b the efficiency of the appliance in part **a** if it requires an input power of 40 W

....

....

c the power required from a car engine to maintain a constant speed of 72 $km\,h^{-1}$ against resistive forces of 8 kN

....

....

10 A water pump designed to move water to a higher reservoir is rated at 300 W and moves 300 kg of water every minute. It lifts the water through a height of 4 m and ejects it at a speed of 3 $m\,s^{-1}$. Find the efficiency of the pump.

....

....

....

....

Topic 6 Bulk properties of solids

Important definitions

A **stiff** sample requires a large tensile force to cause it to stretch. The **stiffness**, k, of a sample is the tensile force required to produce an extension of 1 m. Stiffness is measured in N m^{-1} and depends on three things:

- the material that the sample is made from
- the cross-sectional area of the sample
- the length of the sample

Tensile stress, σ, is the tension per unit cross-sectional area of a sample. It is given by:

$$\sigma = \frac{F}{A}$$

and is measured in N m^{-2} or Pa.

Tensile strain, ε, is the ratio of the extension of a sample to its original length. It is given by:

$$\varepsilon = \frac{e}{l}$$

and has no units.

The **Young modulus**, E, of a material is the ratio of tensile stress to the tensile strain produced. It is given by:

$$E = \frac{\sigma}{\varepsilon} = \frac{F}{A} \times \frac{l}{e}$$

and is measured in N m^{-2} or Pa.

The **ultimate tensile stress** (UTS) of a material is the maximum tensile stress it can endure prior to breaking. The term **strong** is used to describe a material with a high UTS.

If a material undergoes **elastic deformation**, it will return to its original length when the load is removed. If the material deforms **plastically**, it will not return to its original length when the load is removed.

The **limit of proportionality** is the point on a force–extension graph where the relationship between the two variables ceases to be linear.

The **elastic limit** is the point beyond which further deformation will be plastic.

A **ductile** material can be drawn into wires: it will undergo significant plastic deformation prior to breaking.

Malleable materials can be dented and/or hammered into shape.

Brittle materials, when subjected to a tensile force, undergo little or no plastic deformation prior to breaking. Brittle materials are neither ductile nor malleable; they have a low **fracture energy** and therefore snap easily when they are bent. Cracks propagate easily in brittle materials, so such materials have a tendency to shatter rather than break cleanly.

A **hard** material can scratch a material that is **soft**. Hardness is often measured according to the **Mohs hardness scale**, which grades hardness on a scale of 1 (e.g. talc) to 10 (e.g. diamond).

The **density**, ρ, of a material is the mass per unit volume. It is given by:

$$\rho = \frac{M}{V}$$

and has units of kg m^{-3}.

Hooke's law

A **Hookean** material is one that obeys **Hooke's law**:
The **extension** of the material is **directly proportional** to the **tension** in the sample, *up to the limit of proportionality.*

Hooke's law can be summarised by the equation $F = ke$, where F is the tension in the sample, e is the extension of the sample, and k is the **stiffness** of the sample.

Many materials obey Hooke's law, including most metals. The stiffness of a sample can be found by hanging masses from it and recording the extension from the original length. When the sample is in equilibrium, the tension in the sample is equal to the applied weight, which can in turn be calculated using the equation $W = mg$. A **force–extension** graph can then be plotted.

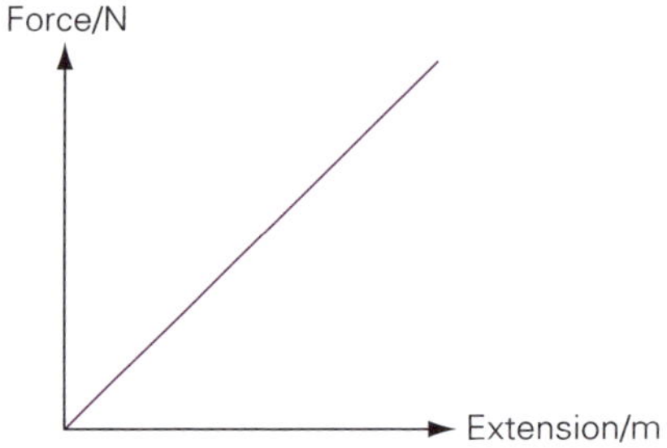

Key point: Usually the independent variable in an investigation is plotted along the *x*-axis of a graph, but a force–extension graph is a rare exception to the rule. The graph is plotted this way around so that the **gradient of the best-fit line** is the **stiffness** of the sample.

The energy stored in the sample at any amount of extension can be found using the equation:

work done = Fs

However, because the tension in the sample does not stay constant as the sample is extended but, rather, varies in direct proportion to the extension, the average force $\frac{1}{2}F$ must be used. The formula therefore becomes:

energy stored = $\frac{1}{2}Fe$

where F is the tension in the sample when the extension is e. That is to say, **the energy stored is given by the area beneath the force–extension graph**. Substituting for F gives us a second equation that can be used when we know only k and e:

energy stored = $\frac{1}{2}ke^2$

Springs

Springs obey Hooke's law; the stiffness, k, of a spring is often referred to as the **spring constant**. The extension of systems of identical springs in series and/or in parallel can be predicted using knowledge of mechanics.

Two springs arranged in parallel will distribute a weight equally between them. They will therefore each extend by half the amount they would have extended if taking the full weight alone.

The same springs connected end-to-end in series will each experience a tension equal to the total weight applied. Therefore they will *each* extend by the full amount.

Hysteresis

Some materials, such as rubber, have a non-linear relationship between force and extension that is different for loading and unloading. The shape of the graph produced is called a **hysteresis curve**. The area beneath the loading curve gives the energy transferred to the sample during loading; the area beneath the unloading curve gives the energy released during unloading. The area enclosed between the loading and unloading curves gives the energy in the sample that gets converted to thermal energy during the loading–unloading cycle.

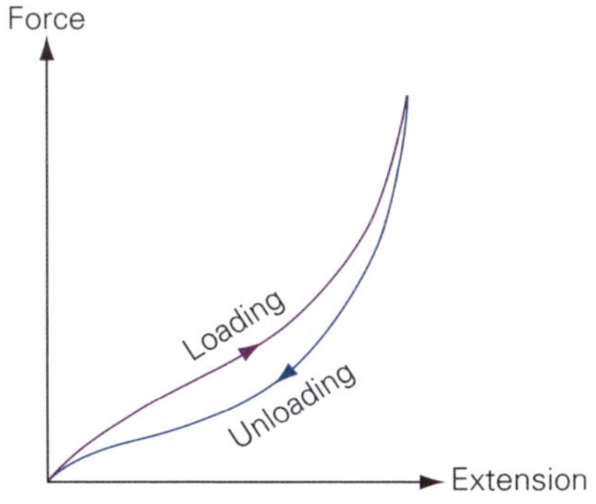

Young modulus

The Young modulus is constant for a given material, regardless of its dimensions. The stiffness of a sample is directly proportional to both the Young modulus of the material and the cross-sectional area of the sample; it is also inversely proportional to the original length of the sample. Thus we have $k \propto E$, $k \propto A$, $k \propto 1/l$, and we can write:

$$k = \frac{EA}{l}$$

The Young modulus of a material can be found by plotting a graph of **tensile stress against tensile strain**; the Young modulus is given by the **gradient of the linear region**.

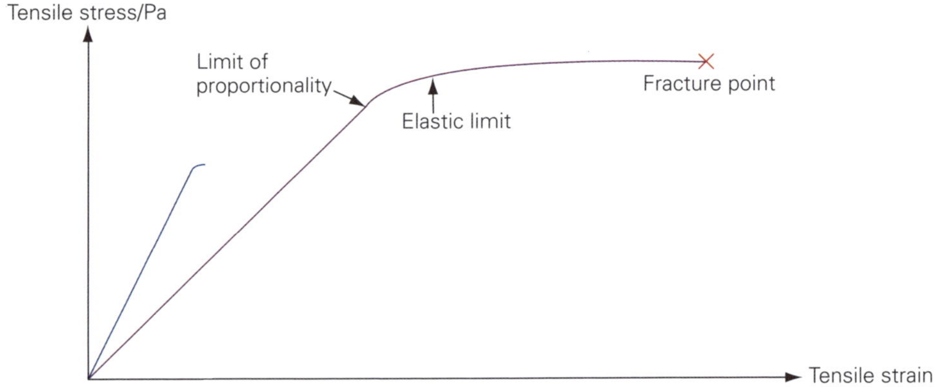

The graph shows two different materials:

- The purple line represents a ductile material.
- The blue line represents a material with a higher Young modulus but which is brittle and has a lower ultimate tensile stress.

It can be shown that the **area beneath the linear portion of a stress–strain graph is the energy stored per unit volume** in the sample:

$$\text{energy stored per unit volume} = \frac{\frac{1}{2}Fe}{A \times l} = \frac{1}{2} \times \frac{F}{A} \times \frac{e}{l} = \frac{1}{2}\sigma\varepsilon$$

A Hookean material will deform elastically in a linear fashion up to the **limit of proportionality**. Within this linear region, work is being done against the attractive forces between the particles; if the load is removed, the sample will return to its original length. Beyond the limit of proportionality, the sample will continue to deform elastically for a short while, until the **elastic limit** is reached. Beyond the elastic limit, plastic deformation will occur rapidly, where rows of atoms slide over each other until the ultimate tensile stress is reached and the sample breaks.

Exam tip: The cross-sectional area of a cylindrical wire is given by $A = \pi r^2$, where r is the radius of the wire, measured in metres. Take care to note whether you are given the diameter or the radius of the wire. It is likely you will be given this dimension in millimetres: it is preferable to convert to SI units before calculating A, as it is easy to forget that there are $10^6\,\text{mm}^2$ in $1\,\text{m}^2$.

Practice questions

1 State the meaning of the following terms, giving equations and/or units where appropriate:

a Hooke's law

b stiffness

c elastic limit

d ductile

e brittle

f ultimate tensile stress

2 Calculate the stiffness of each of the following:

a Sample A, which deforms by 5 mm when a mass of 200 g is hung from it

b Sample B, which deforms by 6.25 μm when a mass of 500 g is hung from it

....

....

....

3 What is the extension of a spring with a spring constant of 120 N m^{-1} if:

a a mass of 40 g is added to it?

....

....

....

b a mass of 800 g is added to it?

....

....

....

4 What **mass** would need to be hung from the following samples in order to produce the following extensions?

a Stiffness of 49 N m^{-1}; extension of 0.6 m.

....

....

....

b Stiffness of 2.45 MN m^{-1}; extension of 3 μm.

....

....

....

5 A student has a bag of ten identical springs, each of which has a spring constant of $50\,N\,m^{-1}$. What would be the extension of each of the following combinations of springs if they were subjected to a tensional force of 5 N? Show all your working.

a A single spring.

....

....

b Two springs in parallel.

....

....

....

c Five springs in parallel.

....

....

d Two springs in series.

....

....

....

e Five springs in series.

....

....

f Five parallel sets of two springs in series.

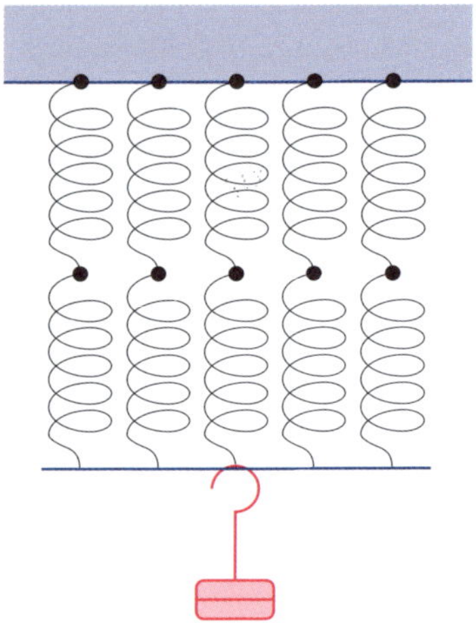

....

....

....

....

6 The table shows some data taken by a student using a Mecmesin tester for a sample of copper.

Weight (N)	Length (cm)	Extension (cm)
0	10.00	
10	10.12	
20	10.24	
30	10.37	
40	10.48	
50	10.60	
60	10.71	
70	10.82	
80	11.00	
90	12.50	

a Complete the last column of the table.

b Plot a force–extension graph on the grid below.

c Add a best-fit line to the graph.

d Label the limit of proportionality.

e Label where the elastic limit might be.

f What is the stiffness of the sample?

g How can the energy stored in the sample be found from the graph?

h Find the energy stored:

i when the extension is 5 mm

ii when the extension is 3 mm

Synoptic questions

1 The particle in the diagram below has a mass of 2 kg and is being pulled up a rough slope by a machine that provides a constant force T of 30.0 N. A frictional force, F, of 5.91 N opposes the motion.

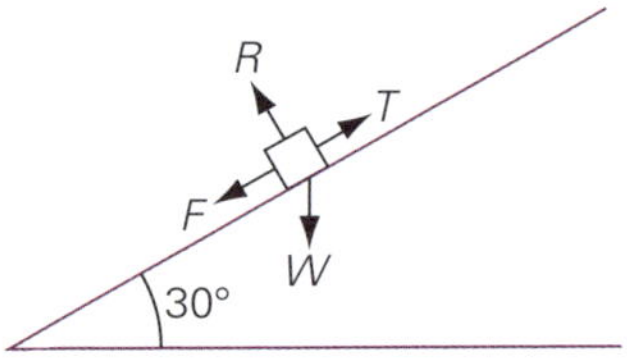

a What is the component of the particle's weight that acts:

i parallel to the slope?

....

....

ii perpendicular to the slope?

....

....

b What is the magnitude of the reaction force R exerted by the slope on the particle?

....

c What is the acceleration of the particle up the slope?

....

....

....

....

d Suppose that in this way the particle is raised through a vertical height of 75 cm in 2.4 s.

i Calculate the work done against the particle's weight.

....

....

ii Calculate the work done against friction.

....

....

....

e The machine that is providing the force *F* draws a current of 40 mA from a 240 V supply. What is the efficiency of the machine?

....

....

....

....

....

2 Using the following information about the horizontal forces acting on a 1500 kg car, sketch a velocity–time graph for the journey in as much detail as you can, on the axes provided. Show all of your working.

t = 0 The car starts from rest.

0–2 s The engine produces uniform thrust of 4.5 kN; frictional forces are negligible.

2–8 s The accelerator pedal is released until the car is travelling at a steady speed of 10 m s^{-1}.

8–12 s The car is in equilibrium.

12–16 s The accelerator pedal is released completely and the brakes are applied; resistive forces total 2625 N.

16–20 s The brake pedal is gradually released, reducing the deceleration until the car comes to rest.

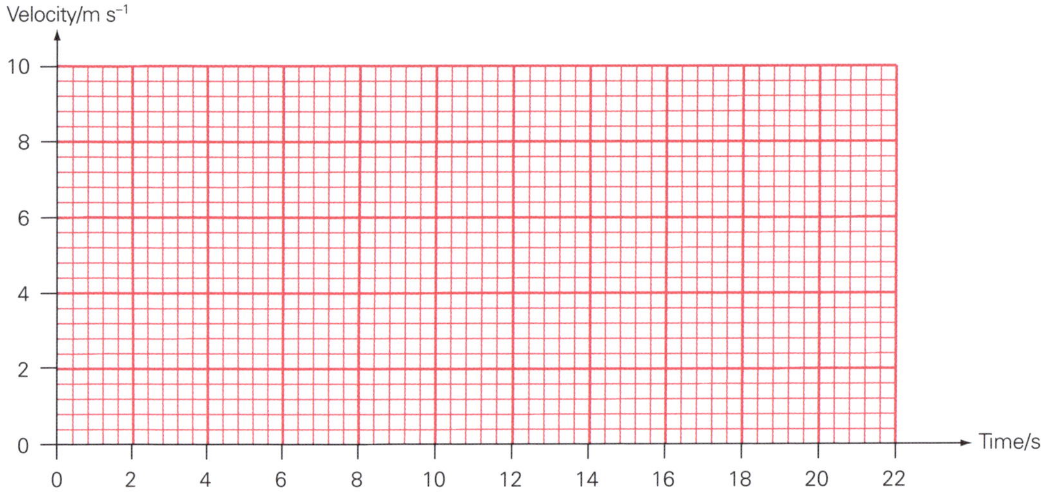
Velocity/m s⁻¹
10
8
6
4
2
0
0
2
4
6
8
10
12
14
16
18
20
22
Time/s

3 Strobe photography (sometimes called multiflash photography) is often used to analyse the motion of projectiles. In a darkened room, a camera shutter is kept open while a stroboscope provides regular flashes of light. Several images of the moving object are therefore recorded at time intervals of Δt; the images are then superimposed on one another in a compound image. The diagram below shows a strobe photograph of two identical balls: ball 1 (on the left) was dropped vertically and ball 2 (on the right) was projected horizontally.

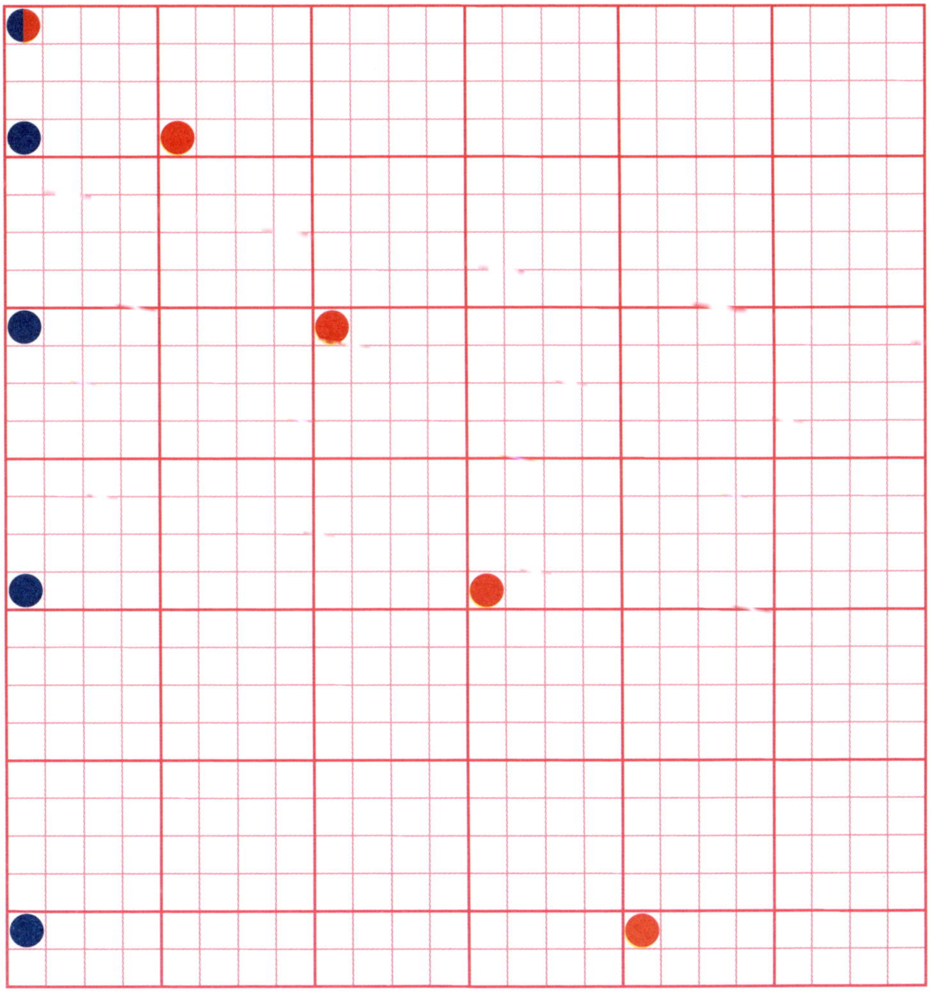

a What word or phrase describes the motion of both balls?

....

b What can you say about the horizontal separation between consecutive images of ball 2? What does this tell you?

....

....

c Write down an equation that could be used to find:

i the height fallen by ball 1 after a time *t*

....

ii the horizontal displacement of ball 2 after a time *t*

....

d Write a method, in bullet points, to explain how strobe photography could be used to find a value for *g*. Explain what measurements you would need to take and the calculations you would perform.

....

....

....

....

....

....

....

....

....

....

....

....

4 A bungee jumper of mass 60 kg jumps off a bridge 100 m above the surface of a river below. The bungee cord's original length is 50 m and its diameter is 1 cm.

a Where would be a suitable place to define zero gravitational potential energy in this situation?

....

b Calculate the jumper's gravitational potential energy when he is standing on the bridge.

....

....

c Assuming that the rope obeys Hooke's law, sketch a diagram to show how the elastic potential energy stored in the rope varies with the height fallen. Label important values.

d Using the law of conservation of energy, calculate the stiffness of the rope if the jumper just avoids entering the water.

Hint: the tension in the rope is not constant as the jumper falls.

....

....

....

....

e Calculate the Young modulus of the rope.

....

....

f When the jumper is 10 m above the water:

i What is the strain in the rope?

....

....

ii What is the stress in the rope?

....

....

....

iii How much elastic potential energy is stored in the rope?

....

....

iv By energy considerations, calculate the jumper's kinetic energy at this point, and hence show that his speed is 23 $m\,s^{-1}$.

....

....

....

....

....